SCIENTIFIC AMERICAN EDUCATIONAL PUBLISHING

SCIENTIFIC AMERICAN INVESTIGATES GEOLOGY

THE ROCK CYCLE

TARON LONGORIA

Published in 2025 by The Rosen Publishing Group
in association with Scientific American Educational Publishing
2544 Clinton Street, Buffalo NY 14224

Library of Congress Cataloging-in-Publication Data

Names: Longoria, Taron, author.
Title: The rock cycle / Taron Longoria.
Description: Buffalo, New York : Scientific American Educational Publishing, an imprint of Rosen Publishing, [2025] | Series: Scientific American investigates geology | Includes bibliographical references and index.
Identifiers: LCCN 2024009884 | ISBN 9781725351097 (library binding) | ISBN 9781725351080 (paperback) | ISBN 9781725351103 (ebook)
Subjects: LCSH: Geochemical cycles–Juvenile literature. | Petrology–Juvenile literature.
Classification: LCC QE432.2 .L66 2025 | DDC 552/.06–dc23/eng/20240402
LC record available at https://lccn.loc.gov/2024009884

Portions of this work were originally authored by Maria Nelson and published as *The Rock Cycle*. All new material in this edition is authored by Taron Longoria.

Designer: Tanya Dellaccio Keeney
Editor: Brianna Propis

Photo credits: Cover www.sandatlas.org/Shutterstock.com; p. 5 Ko Zatu/Shutterstock.com; p. 7 FoxGrafy/Shutterstock.com; p. 9 VectorMine/Shutterstock.com; p. 11 kubais/Shutterstock.com; p. 13 Thomas Dekiere/Shutterstock.com; p. 15 ElenVD. Shutterstock.com; p. 17 Maridav/Shutterstock.com; p. 19 Narongsak Nagadhana/Shutterstock.com; p. 21 PeopleImages.com - Yuri A/Shutterstock.com.

Printed in the United States of America

CPSIA compliance information: Batch #CSSA25. For Further Information contact Rosen Publishing at 1-800-237-9932.

CONTENTS

Words in the glossary appear in **bold** type the first time they are used in the text.

THE CYCLE

Do you ever wonder what Earth is made of? During the 1700s, scientists were very curious about the makeup and age of Earth. They examined rocks and **fossils** but couldn't find answers to their questions. Then, a Scottish scientist named James Hutton presented an idea. He said that the rocks on Earth have been gradually forming, breaking down, and reforming in the same way for many years.

Today, Hutton's idea is known as the rock cycle. A cycle refers to a series of repeating steps over time.

Mountains forming is part of the rock cycle!

FUN FACT

ROCK IS THE NATURALLY OCCURRING, NONLIVING SOLID MATTER THAT MAKES UP EARTH. IT CAN BE MADE OF A SINGLE **MINERAL** OR OF SEVERAL MINERALS THAT ARE COMPRESSED, OR PACKED TIGHTLY TOGETHER.

FOUR LAYERS

Scientists have learned a lot about Earth in the time after James Hutton wrote about **geology** in the 1700s. They discovered that the planet has four main layers: a hard outer crust, a hot mantle, an even hotter, liquid outer core, and a solid inner core.

Together, the crust and solid upper part of the mantle are called the lithosphere. It's divided into moveable slabs of rock called plates. The plates slide past one another, **collide**, and slip under and over each other. The rock cycle primarily happens in the lithosphere.

STRUCTURE OF THE EARTH

FUN FACT

GEOLOGISTS ARE SCIENTISTS WHO STUDY WHAT EARTH IS MADE OF AND HOW IT WAS FORMED.

This illustration shows the main layers of Earth. The part we see—and live on—is called the crust, but it is only a small part of the entire planet.

EARTH'S CRUST

Sedimentary rock makes up most of Earth's crust. This rock forms from the pressing together of tiny bits of rock—such as sand, clay, or pebbles—called sediment. Metamorphic rock forms from existing rocks that change to remain **stable** under new conditions, such as increased temperature. Igneous rock forms from the cooling of hot, melted rock rising from within Earth.

All three kinds of rock are part of the rock cycle. Together, they are a part of an amazing Earth **process**!

FUN FACT

ALL ROCK ORIGINATES, OR BEGINS, AS IGNEOUS ROCK.

THE ROCK CYCLE

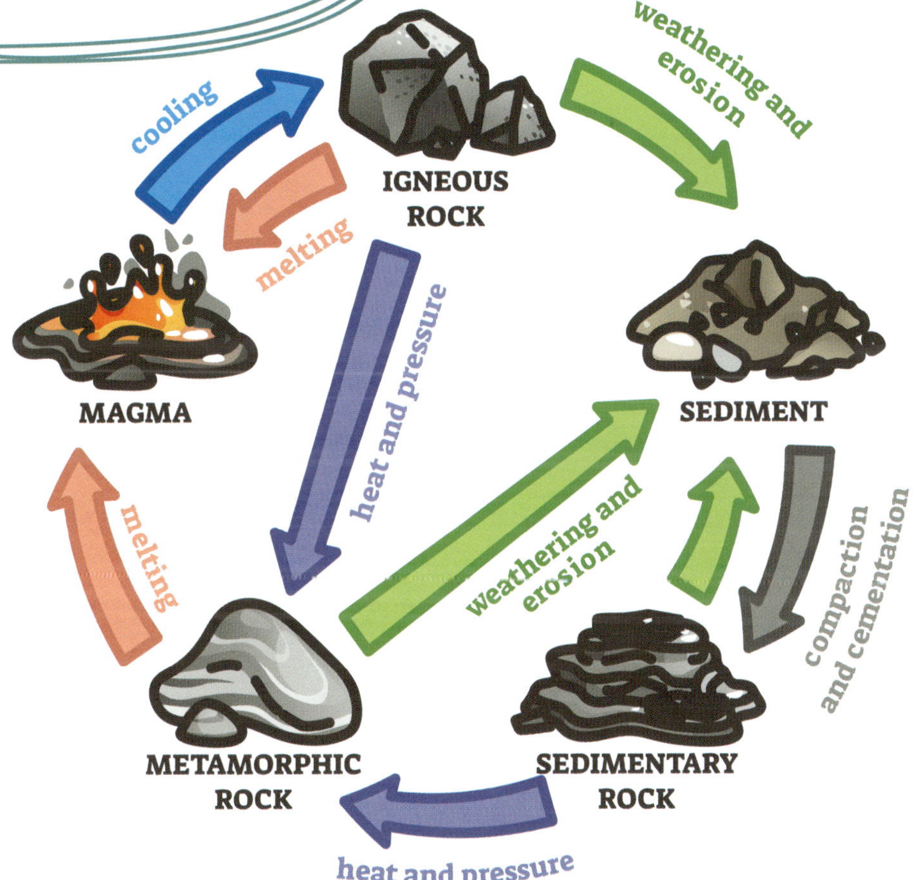

The rock cycle shows how each kind of rock can become either of the other two. With enough heat and **pressure**, a sedimentary rock such as limestone can become a metamorphic rock, like marble.

WEATHERING

The breakdown of rock into sediment—known as weathering—begins the rock cycle. Weathering can happen when water flows into a rock and freezes, causing it to crack. Mixtures of water and minerals can cause **chemical** changes in rock, making it soft enough to fall apart. Wind blowing over rock for a long time can weather it too.

Tiny living things called bacteria give off chemicals that break down rock. Animals can also cause weathering by scratching at or digging into rocks.

MOLE

Weathering that is caused by a living thing is called biological weathering.

FUN FACT

BURROWING ANIMALS, SUCH AS MOLES AND EARTHWORMS, CAUSE WEATHERING. THEY MOVE ROCK **FRAGMENTS** TO THE SURFACE BY DIGGING HOLES IN THE GROUND.

ROCKSLIDES

Sediment created by weathering usually moves. The wearing away of rock into sediment and its movement is called erosion. Wind and water are both common means of erosion.

Gravity can also cause erosion. It pulls loose rock down mountains in rockslides. Rockslides occur in areas with lots of mountains or places that have lots of loose, uncovered rock.

After a time, sediment settles in a place. Then, more sediment settles on top of it.

Erosion like rockslides can cause damage to homes, property, and people!

FUN FACT

EROSION CAN ALSO BE CAUSED BY ICE, USUALLY IN THE FORM OF A **GLACIER**. AS GLACIERS MOVE DOWNHILL OR ACROSS LAND, THEY CAN MOVE EVERYTHING IN THEIR PATH—FROM TINY GRAINS OF SAND TO HUGE ROCKS.

CHEMICAL CHANGES

Pressure increases as more matter buries sediment, causing **compaction**. Water flows into open spaces in the rock, leaving behind minerals that bind the sediment together. This is called cementation. The high pressure plus cementation is how sedimentary rock forms!

Some rock stays in this form. Some continues to be forced deeper into Earth. Heat and pressure increase. These changes around the rock cause its chemical makeup to change in order to stay stable. It becomes metamorphic rock!

Clasts are fragments of rock and other matter that make up sedimentary rock. In some kinds of rock, they can be seen, such as in the rock shown here.

FUN FACT

THE MOVEMENT OF EARTH'S TECTONIC PLATES IS ONE WAY ROCK CAN BE PUSHED BELOW EARTH'S SURFACE. TECTONIC PLATES ARE THE LARGE PIECES OF ROCK THAT MAKE UP EARTH'S OUTER LAYER.

RISING AND COOLING

Rock can also melt. This can be caused by both increased temperature and changes in the pressure on a rock. This liquid rock is called magma when it's under Earth's surface. Some kinds of igneous rock form when magma cools in pockets and tunnels underground.

Magma sometimes bursts out of cracks in Earth's surface or the weak places where plates meet. These are volcanoes! Magma is called lava when it's above the ground. Cooled lava also forms igneous rocks.

FUN FACT

ACTIVE VOLCANOES HAVE A MOUNTAIN SHAPE AROUND THEM. THIS FORMS FROM IGNEOUS ROCK THAT BUILDS UP AROUND THE VOLCANO'S OPENING.

Kīlauea in Hawaii is an example of an active volcano. This picture was taken during an **eruption** in 2018.

UPLIFT

Just as the movement of Earth's plates can force rocks deeper underground, the collision of tectonic plates may raise rock to the surface. This is called uplift—and it's one of the most important parts of the rock cycle.

Uplift pushes rock that formed underground to Earth's surface as mountains and hills. The rock then faces weathering to begin the rock cycle again. Without uplift, the rock cycle would stop over time. The existing rock, mountains, and hills on Earth would wear away!

Earthquakes can occur near Earth's surface or deep below it. Some earthquakes are so small that people don't even feel them, and others can destroy homes, roads, bridges, and more.

FUN FACT

EARTHQUAKES ARE CAUSED BY THE MOVEMENT OF TECTONIC PLATES. THEY MAY HAPPEN WHEN PLATES GET STUCK AND THEN SUDDENLY MOVE PAST EACH OTHER.

AN ONGOING PROCESS

The rock cycle has been occurring since Earth formed—and it's still occurring right now! Even though we can follow its steps, it has no real beginning or end.

Beginning with James Hutton's ideas, scientists continue to study how Earth formed and keeps changing. They have learned how to figure out how old a rock is and what kind of rock they've found by studying its features and chemical makeup. Because of the work of scientists, we now know that Earth is about 4.5 billion years old. That's a lot of ancient rock!

By studying rocks, scientists have learned about ancient life and life-forms on Earth.

GLOSSARY

burrowing: Making a hole or tunnel in the ground.

chemical: Having to do with matter that can be mixed with other matter to cause changes. Also, the matter itself.

collide: The hitting of two objects against each other.

compaction: Forcing things closer together.

eruption: The bursting forth of molten rock, steam, and lava from a volcano.

fossil: The marks or remains of plants and animals that lived long ago.

fragment: A small part broken off from a whole.

geology: The science that studies the history of Earth and its life as recorded in rocks.

glacier: A large body of ice that moves slowly.

gravity: The force that pulls objects toward Earth's center.

mineral: Nonliving matter that make up Earth's rocks, sands, and soils.

pressure: The application of force.

process: The set of steps that move something forward.

stable: Not likely to change suddenly or greatly.

FOR MORE INFORMATION

Books

Carlson-Berne, Emma. *Let's Explore the Rock Cycle.* Minneapolis, MN: Lerner Publications, 2021.

Gieseke, Tyler. *The Rock Cycle.* Minneapolis, MN: DiscoverRoo, an imprint of Pop!, 2023.

Websites

Britannica Kids: Erosion
kids.britannica.com/kids/article/erosion/399447
Learn more about erosion, including the different types and the damage it can cause.

National Geographic: The Rock Cycle
education.nationalgeographic.org/resource/rock-cycle/
Read more about the amazing process of the rock cycle.

Publisher's note to educators and parents: Our editors have carefully reviewed these websites to ensure that they are suitable for students. Many websites change frequently, however, and we cannot guarantee that a site's future contents will continue to meet our high standards of quality and educational value. Be advised that students should be closely supervised whenever they access the internet.

INDEX